DARWINISM APPLIED TO MAN

(1889)

BY

ALFRED RUSSEL WALLACE

British Library Cataloguing-in-Publication Data
A catalogue record for this book is available from the
British Library

Alfred Russel Wallace

Alfred Russel Wallace was born on 8th January 1823 in the village of Llanbadoc, in Monmouthshire, Wales.

At the age of five, Wallace's family moved to Hertford where he later enrolled at Hertford Grammar School. He was educated there until financial difficulties forced his family to withdraw him in 1836. He then boarded with his older brother John before becoming an apprentice to his eldest brother, William, a surveyor. He worked for William for six years until the business declined due to difficult economic conditions.

After a brief period of unemployment, he was hired as a master at the Collegiate School in Leicester to teach drawing, map-making, and surveying. During this time he met the entomologist Henry Bates who inspired Wallace to begin collecting insects. He and bates continued exchanging letters after Wallace left teaching to pursue his surveying career. They corresponded on prominent works of the time such as Charles Darwin's *The Voyage of the Beagle* (1839) and Robert Chamber's *Vestiges of the Natural History of Creation* (1844).

Wallace was inspired by the travelling naturalists of the day and decided to begin his exploration career collecting specimens in the Amazon rainforest. He explored the Rio Negra for four years, making notes on the peoples and languages he encountered as well as the geography, flora, and fauna. On his return voyage his ship, Helen, caught fire and he and the crew were stranded for ten days before being picked up by the Jordeson, a brig travelling from Cuba to London. All of his specimens aboard Helen had been lost.

After a brief stay in England he embarked on a journey to the Malay Archipelago (now Singapore, Malaysia, and Indonesia). During this eight year period he collected more than 126,000 specimens, several thousand of which represented new species to science. While travelling, Wallace refined his thoughts about evolution and in 1858 he outlined his theory of natural selection in an article he sent to Charles Darwin. This was published in the same year along with Darwin's own theory. Wallace eventually published an account of his travels *The Malay Archipelago* in 1869, and it became one of the most popular books of scientific exploration in the 19th century.

Upon his return to England, in 1862, Wallace became a staunch defender of Darwin's landmark work *On the Origin of Species* (1859). He wrote responses to those critical of the theory of natural selection, including 'Remarks on the Rev. S. Haughton's Paper on the Bee's Cell, And on the Origin of Species' (1863) and 'Creation by Law' (1867). The former of these was particularly pleasing to Darwin. Wallace also published important papers such as 'The Origin of Human Races and the Antiquity of Man Deduced from the Theory of 'Natural Selection" (1864) and books, including the much cited *Darwinism* (1889).

Wallace made a huge contribution to the natural sciences and he will continue to be remembered as one of the key figures in the development of evolutionary theory.

Wallace died on 7th November 1913 at the age of 90. He is buried in a small cemetery at Broadstone, Dorset, England.

DARWINISM APPLIED TO MAN

General identity of human and animal structure--
Rudiments and variations showing relation of man to other
mammals--The embryonic development of man and other
mammalia--Diseases common to man and the lower animals-
-The animals most nearly allied to man--The brains of man
and apes--External differences of man and apes--Summary of
the animal characteristics of man--The geological antiquity
of man--The probable birthplace of man--The origin of the
moral and intellectual nature of man--The argument from
continuity--The origin of the mathematical faculty--The
origin of the musical and artistic faculties--Independent
proof that these faculties have not been developed by natural
selection--The interpretation of the facts--Concluding
remarks.

Our review of modern Darwinism might fitly have
terminated with the preceding chapter; but the immense
interest that attaches to the origin of the human race, and
the amount of misconception which prevails regarding the
essential teachings of Darwin's theory on this question, as
well as regarding my own special views upon it, induce me

to devote a final chapter to its discussion.

To any one who considers the structure of man's body, even in the most superficial manner, it must be evident that it is the body of an animal, differing greatly, it is true, from the bodies of all other animals, but agreeing with them in all essential features. The bony structure of man classes him as a vertebrate; the mode of suckling his young classes him as a mammal; his blood, his muscles, and his nerves, the structure of his heart with its veins and arteries, his lungs and his whole respiratory and circulatory systems, all closely correspond to those of other mammals, and are often almost identical with them. He possesses the same number of limbs terminating in the same number of digits as belong fundamentally to the mammalian class. His senses are identical with theirs, and his organs of sense are the same in number and occupy the same relative position. Every detail of structure which is common to the mammalia as a class is found also in man, while he only differs from them in such ways and degrees as the various species or groups of mammals differ from each other. If, then, we have good reason to believe that every existing group of mammalia has descended from some common ancestral form--as we saw to be so completely demonstrated in the case of the horse tribe,--and that each family, each order, and even the whole class must similarly have descended from some much more ancient and more generalised type, it would be in the highest degree improbable--so improbable as to be almost inconceivable--that man, agreeing with them so closely in every detail of his structure, should have had some quite distinct mode of origin. Let us, then, see what other evidence bears upon the question, and whether it is sufficient to convert the probability of his animal origin into a practical certainty.

Rudiments and Variations as Indicating the Relation of Man to other Mammals.

All the higher animals present rudiments of organs which, though useless to them, are useful in some allied group, and are believed to have descended from a common ancestor in which they were useful. Thus there are in ruminants rudiments of incisor teeth which, in some species, never cut through the gums; many lizards have external rudimentary legs; while many birds, as the Apteryx, have quite rudimentary wings. Now man possesses similar rudiments, sometimes constantly, sometimes only occasionally present, which serve intimately to connect his bodily structure with that of the lower animals. Many animals, for example, have a special muscle for moving or twitching the skin. In man there are remnants of this in certain parts of the body, especially in the forehead, enabling us to raise our eyebrows; but some persons have it in other parts. A few persons are able to move the whole scalp so as to throw off any object placed on the head, and this property has been proved, in one case, to be inherited. In the outer fold of the ear there is sometimes a projecting point, corresponding in position to the pointed ear of many animals, and believed to be a rudiment of it. In the alimentary canal there is a rudiment--the vermiform appendage of the cæcum--which is not only useless, but is sometimes a cause of disease and death in man; yet in many vegetable feeding animals it is very long, and even in the orang-utan it is of considerable length and convoluted. So, man possesses rudimentary bones of a tail concealed beneath the skin, and, in some rare cases, this forms a minute external tail.

The variability of every part of man's structure is very great, and many of these variations tend to approximate towards the structure of other animals. The courses of the arteries are eminently variable, so that for surgical purposes it has been necessary to determine the probable proportion of each variation. The muscles are so variable that in fifty cases the muscles of the foot were found to be not strictly alike in any two, and in some the deviations were considerable; while in thirty-six subjects Mr. J. Wood observed no fewer than 558 muscular variations. The same author states that in a single male subject there were no fewer than seven muscular variations, all of which plainly represented muscles proper to various kinds of apes. The muscles of the hands and arms-

-parts which are so eminently characteristic of man--are extremely liable to vary, so as to resemble the corresponding muscles of the lower animals. That such variations are due to reversion to a former state of existence Mr. Darwin thinks highly probable, and he adds: "It is quite incredible that a man should, through mere accident, abnormally resemble certain apes in no less than seven of his muscles, if there had been no genetic connection between them. On the other hand, if man is descended from some ape-like creature, no valid reason can be assigned why certain muscles should not suddenly reappear after an interval of many thousand generations, in the same manner as, with horses, asses, and mules, dark coloured stripes suddenly reappear on the legs and shoulders, after an interval of hundreds, or more probably of thousands of generations."[1]

The Embryonic Development of Man and other Mammalia.

The progressive development of any vertebrate from the ovum or minute embryonic egg affords one of the most marvellous chapters in Natural History. We see the contents of the ovum undergoing numerous definite changes, its interior dividing and subdividing till it consists of a mass of cells, then a groove appears marking out the median line or vertebral column of the future animal, and thereafter are slowly developed the various essential organs of the body. After describing in some detail what takes place in the case of the ovum of the dog, Professor Huxley continues: "The history of the development of any other vertebrate animal, lizard, snake, frog, or fish tells the same story. There is always to begin with, an egg having the same essential structure as that of the dog; the yolk of that egg undergoes division or segmentation, as it is called, the ultimate products of that segmentation constitute the building materials for the body of the young animal; and this is built up round a primitive groove, in the floor of which a notochord is developed. Furthermore, there is a period in which the young of all these animals resemble one another, not merely in outward form, but in all essentials of structure, so closely, that the differences between them are inconsiderable, while in their subsequent course they diverge more and more widely from one another. And it is a general law that the more closely any animals resemble one another in adult structure, the larger and the more intimately do their embryos resemble one another; so that, for example, the embryos of a snake and of a lizard remain like one another longer than do those of a snake and a bird; and the embryos of a dog and of a cat remain like one another for a far longer period than do those of a dog and a bird, or of a dog and an opossum, or even than those of a dog and a monkey."[2]

We thus see that the study of development affords a test of affinity in animals that are externally very much unlike each other; and we naturally ask how this applies to man. Is he developed in a different way from other mammals, as we should certainly expect if he has had a distinct and altogether different origin? "The reply," says Professor Huxley, "is not doubtful for a moment. Without question, the mode of origin and the early stages of development of man are identical with those of the animals immediately below him in the scale." And again he tells us: "It is very long before the body of the young human being can be readily discriminated from that of the young puppy; but at a tolerably early period the two become distinguishable by the different forms of their adjuncts, the yelk-sac and the allantois;" and after describing these differences he continues: "But exactly in those respects in which the developing man differs from the dog, he resembles the ape. . . . So that it is only quite in the latter stages of development that the young human being presents marked differences from the young ape, while the latter departs as much from the dog in its development as the man does. Startling as this last assertion may appear to be, it is demonstrably true, and it alone appears to me sufficient to place beyond all doubt the structural unity of man with the rest of the animal world, and more particularly and closely with the apes."[3]

A few of the curious details in which man passes through

stages common to the lower animals may be mentioned. At one stage the os coccyx projects like a true tail, extending considerably beyond the rudimentary legs. In the seventh month the convolutions of the brain resemble those of an adult baboon. The great toe, so characteristic of man, forming the fulcrum which most assists him in standing erect, in an early stage of the embryo is much shorter than the other toes, and instead of being parallel with them, projects at an angle from the side of the foot, thus corresponding with its permanent condition in the quadrumana. Numerous other examples might be quoted, all illustrating the same general law.

Diseases Common to Man and the Lower Animals.

Though the fact is so well known, it is certainly one of profound significance that many animal diseases can be communicated to man, since it shows similarity, if not identity, in the minute structure of the tissues, the nature of the blood, the nerves, and the brain. Such diseases as hydrophobia, variola, the glanders, cholera, herpes, etc., can be transmitted from animals to man or the reverse; while monkeys are liable to many of the same non-contagious diseases as we are. Rengger, who carefully observed the common monkey (Cebus Azaræ) in Paraguay, found it liable to catarrh, with the usual symptoms, terminating sometimes in consumption. These monkeys also suffered from apoplexy, inflammation of the bowels, and cataract in the eye. Medicines produced the same effect upon them as upon us. Many kinds of monkeys have a strong taste for tea, coffee, spirits, and even tobacco. These facts show the similarity of the nerves of taste in monkeys and in ourselves, and that their whole nervous system is affected in a similar way. Even the parasites, both external and internal, that affect man are not altogether peculiar to him, but belong to the same families or genera as those which infest animals, and in one case, scabies, even the same species.[4] These curious facts seem quite inconsistent with the idea that man's bodily structure and nature are altogether distinct from those of animals, and have had a different origin; while the facts are just what we should expect if he has been produced by descent with modification from some common ancestor.

The Animals most nearly Allied to Man.

By universal consent we see in the monkey tribe a caricature of humanity. Their faces, their hands, their actions and expressions present ludicrous resemblances to our own. But there is one group of this great tribe in which this resemblance is greatest, and they have hence been called the anthropoid or man-like apes. These are few in number, and inhabit only the equatorial regions of Africa and Asia, countries where the climate is most uniform, the forests densest, and the supply of fruit abundant throughout the year. These animals are now comparatively well known, consisting of the orang-utan of Borneo and Sumatra, the chimpanzee and the gorilla of West Africa, and the group of gibbons or long-armed apes, consisting of many species and inhabiting South-Eastern Asia and the larger Malay Islands. These last are far less like man than the other three, one or other of which has at various times been claimed to be the most man-like of the apes and our nearest relations in the animal kingdom. The question of the degree of resemblance of these animals to ourselves is one of great interest, leading, as it does, to some important conclusions as to our origin and geological antiquity, and we will therefore briefly consider it.

If we compare the skeletons of the orang or chimpanzee with that of man, we find them to be a kind of distorted copy, every bone corresponding (with very few exceptions), but altered somewhat in size, proportions, and position. So great is this resemblance that it led Professor Owen to remark: "I cannot shut my eyes to the significance of that all-pervading similitude of structure--every tooth, every bone, strictly homologous--which makes the determination of the difference between *Homo* and *Pithecus* the anatomist's difficulty."

The actual differences in the skeletons of these apes and that of man--that is, differences dependent on the presence or absence of certain bones, and not on their form or position--have been enumerated by Mr. Mivart as follows:--(1) In the breast-bone consisting of but two bones, man agrees with the gibbons; the chimpanzee and gorilla having this part consisting of seven bones in a single series, while in the orang they are arranged in a double series of ten bones. (2) The normal number of the ribs in the orang and some gibbons is twelve pairs, as in man, while in the chimpanzee and gorilla there are thirteen pairs. (3) The orang and the gibbons also agree with man in having five lumbar vertebræ, while in the gorilla and the chimpanzee there are but four, and sometimes only three. (4) The gorilla and chimpanzee agree with man in having eight small bones in the wrist, while the orang and the gibbons, as well as all other monkeys, have nine.[5]

The differences in the form, size, and attachments of the various bones, muscles, and other organs of these apes and man are very numerous and exceedingly complex, sometimes one species, sometimes another agreeing most nearly with ourselves, thus presenting a tangled web of affinities which it is very difficult to unravel. Estimated by the skeleton alone, the chimpanzee and gorilla seem nearer to man than the orang, which last is also inferior as presenting certain aberrations in the muscles. In the form of the ear the gorilla is more human than any other ape, while in the tongue the orang is the more man-like. In the stomach and liver the gibbons approach nearest to man, then come the orang and chimpanzee, while the gorilla has a degraded liver more resembling that of the lower monkeys and baboons.

The Brains of Man and Apes.

We come now to that part of his organisation in which man is so much higher than all the lower animals--the brain; and here, Mr. Mivart informs us, the orang stands highest in rank. The height of the orang's cerebrum in front is greater in proportion than in either the chimpanzee or the gorilla. "On comparing the brain of man with the brains of the orang, chimpanzee, and baboon, we find a successive decrease in the frontal lobe, and a successive and very great increase in the relative size of the occipital lobe. Concomitantly with this increase and decrease, certain folds of brain substance, called 'bridging convolutions,' which in man are conspicuously interposed between the parietal and occipital lobes, seem as utterly to disappear in the chimpanzee, as they do in the baboon. In the orang, however, though much reduced, they are still to be distinguished. . . . The actual and absolute mass of the brain is, however, slightly greater in the chimpanzee than in the orang, as is the relative vertical extent of the middle part of the cerebrum, although, as already stated, the frontal portion is higher in the orang; while, according to M. Gratiolet, the gorilla is not only inferior to the orang in cerebral development, but even to his smaller African congener, the chimpanzee."[6]

On the whole, then, we find that no one of the great apes can be positively asserted to be nearest to man in structure. Each of them approaches him in certain characteristics, while in others it is widely removed, giving the idea, so consonant with the theory of evolution as developed by Darwin, that all are derived from a common ancestor, from which the existing anthropoid apes as well as man have diverged. When, however, we turn from the details of anatomy to peculiarities of external form and motions, we find that, in a variety of characters, all these apes resemble each other and differ from man, so that we may fairly say that, while they have diverged somewhat from each other, they have diverged much more widely from ourselves. Let us briefly enumerate some of these differences.

External Differences of Man and Apes.

All apes have large canine teeth, while in man these are no longer than the adjacent incisors or premolars, the whole forming a perfectly even series. In apes the arms are proportionately much longer than in man, while the thighs are much shorter. No ape stands really erect, a posture which is natural in man. The thumb is proportionately larger in man, and more perfectly opposable than in that of any ape. The foot of man differs largely from that of all apes, in the horizontal sole, the projecting heel, the short toes, and the powerful great toe firmly attached parallel to the other toes; all perfectly adapted for maintaining the erect posture, and for free motion without any aid from the arms or hands. In apes the foot is formed almost exactly like our hand, with a large thumb-like great toe quite free from the other toes, and so articulated as to be opposable to them; forming with the long finger-like toes a perfect grasping hand. The sole cannot be placed horizontally on the ground; but when standing on a level surface the animal rests on the outer edge of the foot with the finger and thumb-like toes partly closed, while the hands are placed on the ground resting on the knuckles. The illustration on the next page (Fig. 37) shows, fairly well, the peculiarities of the hands and feet of the chimpanzee, and their marked differences, both in form and use, from those of man.

The four limbs, with the peculiarly formed feet and hands, are those of arboreal animals which only occasionally and awkwardly move on level ground. The arms are used in progression equally with the feet, and the hands are only adapted for uses similar to those of our hands when the animal is at rest, and then but clumsily. Lastly, the apes are all hairy animals, like the majority of other mammals, man alone having a smooth and almost naked skin. These numerous and striking differences, even more than those of the skeleton and internal anatomy, point to an enormously remote epoch when the race that was ultimately to develop into man diverged from that other stock which continued the animal type and ultimately produced the existing varieties of anthropoid apes.

Fig. 37.—Chimpanzee (Troglodytes niger).

Summary of the Animal Characteristics of Man.

The facts now very briefly summarised amount almost to a demonstration that man, in his bodily structure, has been derived from the lower animals, of which he is the culminating development. In his possession of rudimentary structures which are functional in some of the mammalia; in the numerous variations of his muscles and other organs agreeing with characters which are constant in some apes; in his embryonic development, absolutely identical in character with that of mammalia in general, and closely resembling in its details that of the higher quadrumana; in the diseases which he has in common with other mammalia; and in the wonderful approximation of his skeleton to those of one or other of the anthropoid apes, we have an amount of evidence in this direction which it seems impossible to explain away. And this evidence will appear more forcible if we consider for a moment what the rejection of it implies. For the only alternative supposition is, that man has been specially created--that is to say, has been produced in some quite different way from other animals and altogether independently of them. But in that case the rudimentary structures, the animal-like variations, the identical course of development, and all the other animal characteristics he possesses are deceptive, and inevitably lead us, as thinking beings making use of the reason which is our noblest and most distinctive feature, into gross error.

We cannot believe, however, that a careful study of the facts of nature leads to conclusions directly opposed to the truth; and, as we seek in vain, in our physical structure and the course of its development, for any indication of an origin independent of the rest of the animal world, we are compelled to reject the idea of "special creation" for man, as being entirely unsupported by facts as well as in the highest degree improbable.

The Geological Antiquity of Man.

The evidence we now possess of the exact nature of the resemblance of man to the various species of anthropoid apes, shows us that he has little special affinity for any one rather than another species, while he differs from them all in several important characters in which they agree with each other. The conclusion to be drawn from these facts is, that his points of affinity connect him with the whole group, while his special peculiarities equally separate him from the whole group, and that he must, therefore, have diverged from the common ancestral form before the existing types of anthropoid apes had diverged from each other. Now, this divergence almost certainly took place as early as the Miocene period, because in the Upper Miocene deposits of Western

Europe remains of two species of ape have been found allied to the gibbons, one of them, Dryopithecus, nearly as large as a man, and believed by M. Lartet to have approached man in its dentition more than the existing apes. We seem hardly, therefore, to have reached, in the Upper Miocene, the epoch of the common ancestor of man and the anthropoids.

The evidence of the antiquity of man himself is also scanty, and takes us but very little way back into the past. We have clear proof of his existence in Europe in the latter stages of the glacial epoch, with many indications of his presence in interglacial or even pre-glacial times; while both the actual remains and the works of man found in the auriferous gravels of California deep under lava-flows of Pliocene age, show that he existed in the New World at least as early as in the Old.[7] These earliest remains of man have been received with doubt, and even with ridicule, as if there were some extreme improbability in them. But, in point of fact, the wonder is that human remains have not been found more frequently in pre-glacial deposits. Referring to the most ancient fossil remains found in Europe--the Engis and Neanderthal crania,--Professor Huxley makes the following weighty remark: "In conclusion, I may say, that the fossil remains of Man hitherto discovered do not seem to me to take us appreciably nearer to that lower pithecoid form, by the modification of which he has, probably, become what he is." The Californian remains and works of art, above referred to, give no indication of a specially low form of man; and it remains an unsolved problem why no traces of the long line of man's ancestors, back to the remote period when he first branched off from the pithecoid type, have yet been discovered.

It has been objected by some writers--notably by Professor Boyd Dawkins--that man did not probably exist in Pliocene times, because almost all the known mammalia of that epoch are distinct species from those now living on the earth, and that the same changes of the environment which led to the modification of other mammalian species would also have led to a change in man. But this argument overlooks the fact that man differs essentially from all other mammals in this respect, that whereas any important adaptation to new conditions can be effected in them only by a change in bodily structure, man is able to adapt himself to much greater changes of conditions by a mental development leading him to the use of fire, of tools, of clothing, of improved dwellings, of nets and snares, and of agriculture. By the help of these, without any change whatever in his bodily structure, he has been able to spread over and occupy the whole earth; to dwell securely in forest, plain, or mountain; to inhabit alike the burning desert or the arctic wastes; to cope with every kind of wild beast, and to provide himself with food in districts where, as an animal trusting to nature's unaided productions, he would have starved.[8]

It follows, therefore, that from the time when the ancestral man first walked erect, with hands freed from any active part in locomotion, and when his brain-power became sufficient to cause him to use his hands in making weapons and tools, houses and clothing, to use fire for cooking, and to plant seeds or roots to supply himself with stores of food, the power of natural selection would cease to act in producing modifications of his body, but would continuously advance his mind through the development of its organ, the brain. Hence man may have become truly man--the species, Homo sapiens--even in the Miocene period; and while all other mammals were becoming modified from age to age under the influence of ever-changing physical and biological conditions, he would be advancing mainly in intelligence, but perhaps also in stature, and by that advance alone would be able to maintain himself as the master of all other animals and as the most widespread occupier of the earth. It is quite in accordance with this view that we find the most pronounced distinction between man and the anthropoid apes in the size and complexity of his brain. Thus, Professor Huxley tells us that "it may be doubted whether a healthy human adult brain ever weighed less than 31 or 32 ounces, or that the heaviest gorilla brain has exceeded 20 ounces," although "a full-grown gorilla is probably pretty nearly twice as heavy as a Bosjes man, or as many an European woman."[9] The average human brain, however, weighs 48 or 49 ounces, and if we take the average ape brain at only 2 ounces less than the very largest gorilla's brain, or 18 ounces, we shall see better the enormous increase which has taken place in the brain of man since the time when he branched off from the apes; and this increase will be still greater if we consider that the brains of apes, like those of all other mammals, have also increased from earlier to later geological times.

If these various considerations are taken into account, we must conclude that the essential features of man's structure as compared with that of apes--his erect posture and free hands--were acquired at a comparatively early period, and were, in fact, the characteristics which gave him his superiority over other mammals, and started him on the line of development which has led to his conquest of the world. But during this long and steady development of brain and intellect, mankind must have continuously increased in numbers and in the area which they occupied--they must have formed what Darwin terms a "dominant race." For had they been few in numbers and confined to a limited area, they could hardly have successfully struggled against the numerous fierce carnivora of that period, and against those adverse influences which led to the extinction of so many more powerful animals. A large population spread over an extensive area is also needed to supply an adequate number of brain variations for man's progressive improvement. But this large population and long-continued development in a single line of advance renders it the more difficult to account for the complete absence of human or pre-human remains in all those deposits which have furnished, in such rich abundance, the remains of other land animals. It is true that the remains of apes are also very rare, and we may well suppose that the superior intelligence of man led him to avoid that extensive destruction by flood or in morass which seems to have often overwhelmed other animals. Yet, when we consider that, even in our own day, men are not unfrequently overwhelmed by volcanic eruptions, as in Java and Japan, or carried away in vast numbers by floods, as in Bengal and China, it seems impossible but that ample remains of Miocene and Pliocene man do exist buried in the most recent layers of the earth's crust, and that more extended research or some fortunate discovery will some day bring them to light.

The Probable Birthplace of Man.

It has usually been considered that the ancestral form of man originated in the tropics, where vegetation is most abundant and the climate most equable. But there are some important objections to this view. The anthropoid apes, as well as most of the monkey tribe, are essentially arboreal in their structure, whereas the great distinctive character of man is his special adaptation to terrestrial locomotion. We can hardly suppose, therefore, that he originated in a forest region, where fruits to be obtained by climbing are the chief vegetable food. It is more probable that he began his existence on the open plains or high plateaux of the temperate or subtropical zone, where the seeds of indigenous cereals and numerous herbivora, rodents, and game-birds, with fishes and molluscs in the lakes, rivers, and seas supplied him with an abundance of varied food. In such a region he would develop skill as a hunter, trapper, or fisherman, and later as a herdsman and cultivator,--a succession of which we find indications in the palæolithic and neolithic races of Europe.

In seeking to determine the particular areas in which his earliest traces are likely to be found, we are restricted to some portion of the Eastern hemisphere, where alone the anthropoid apes exist, or have apparently ever existed.

There is good reason to believe, also, that Africa must be excluded, because it is known to have been separated from the northern continent in early tertiary times, and to have acquired its existing fauna of the higher mammalia by a later union with that continent after the separation from it of Madagascar, an island which has preserved for us a sample, as it were, of the early African mammalian fauna, from which not only the anthropoid apes, but all the higher quadrumana are absent.[10] There remains only the great Euro-Asiatic continent; and its enormous plateaux, extending from

Persia right across Tibet and Siberia to Manchuria, afford an area, some part or other of which probably offered suitable conditions, in late Miocene or early Pliocene times, for the development of ancestral man.

It is in this area that we still find that type of mankind--the Mongolian--which retains a colour of the skin midway between the black or brown-black of the negro, and the ruddy or olive-white of the Caucasian types, a colour which still prevails over all Northern Asia, over the American continents, and over much of Polynesia. From this primary tint arose, under the influence of varied conditions, and probably in correlation with constitutional changes adapted to peculiar climates, the varied tints which still exist among mankind. If the reasoning by which this conclusion is reached be sound, and all the earlier stages of man's development from an animal form occurred in the area now indicated, we can better understand how it is that we have as yet met with no traces of the missing links, or even of man's existence during late tertiary times, because no part of the world is so entirely unexplored by the geologist as this very region. The area in question is sufficiently extensive and varied to admit of primeval man having attained to a considerable population, and having developed his full human characteristics, both physical and mental, before there was any need for him to migrate beyond its limits. One of his earliest important migrations was probably into

Africa, where, spreading westward, he became modified in colour and hair in correlation with physiological changes adapting him to the climate of the equatorial lowlands. Spreading north-westward into Europe the moist and cool climate led to a modification of an opposite character, and thus may have arisen the three great human types which still exist. Somewhat later, probably, he spread eastward into North-West America and soon scattered himself over the whole continent; and all this may well have occurred in early or middle Pliocene times. Thereafter, at very long intervals, successive waves of migration carried him into every part of the habitable world, and by conquest and intermixture led ultimately to that puzzling gradation of types which the ethnologist in vain seeks to unravel.

The Origin of the Moral and Intellectual Nature of Man.

From the foregoing discussion it will be seen that I fully accept Mr. Darwin's conclusion as to the essential identity of man's bodily structure with that of the higher mammalia, and his descent from some ancestral form common to man and the anthropoid apes. The evidence of such descent appears to me to be overwhelming and conclusive. Again, as to the cause and method of such descent and modification, we may admit, at all events provisionally, that the laws of variation and natural selection, acting through the struggle for existence and the continual need of more perfect adaptation to the physical and biological environments, may have brought about, first that perfection of bodily structure in which he is so far above all other animals, and in co-ordination with it the larger and more developed brain, by means of which he has been able to utilise that structure in the more and more complete subjection of the whole animal and vegetable kingdoms to his service.

But this is only the beginning of Mr. Darwin's work, since he goes on to discuss the moral nature and mental faculties of man, and derives these too by gradual modification and development from the lower animals. Although, perhaps, nowhere distinctly formulated, his whole argument tends to the conclusion that man's entire nature and all his faculties, whether moral, intellectual, or spiritual, have been derived from their rudiments in the lower animals, in the same manner and by the action of the same general laws as his physical structure has been derived. As this conclusion appears to me not to be supported by adequate evidence, and to be directly opposed to many well-ascertained facts, I propose to devote a brief space to its discussion.

The Argument from Continuity.

Mr. Darwin's mode of argument consists in showing that the rudiments of most, if not of all, the mental and moral faculties of man can be detected in some animals. The manifestations of intelligence, amounting in some cases to distinct acts of reasoning, in many animals, are adduced as exhibiting in a much less degree the intelligence and reason of man. Instances of curiosity, imitation, attention, wonder, and memory are given; while examples are also adduced which may be interpreted as proving that animals exhibit kindness to their fellows, or manifest pride, contempt, and shame. Some are said to have the rudiments of language, because they utter several different sounds, each of which has a definite meaning to their fellows or to their young; others the rudiments of arithmetic, because they seem to count and remember up to three, four, or even five. A sense of beauty is imputed to them on account of their own bright colours or the use of coloured objects in their nests; while dogs, cats, and horses are said to have imagination, because they appear to be disturbed by dreams. Even some distant approach to the rudiments of religion is said to be found in the deep love and complete submission of a dog to his master.[11]

Turning from animals to man, it is shown that in the lowest savages many of these faculties are very little advanced from the condition in which they appear in the higher animals; while others, although fairly well exhibited, are yet greatly inferior to the point of development they have reached in civilised races. In particular, the moral sense is said to have been developed from the social instincts of savages, and to depend mainly on the enduring discomfort produced by any action which excites the general disapproval of the tribe. Thus, every act of an individual which is believed to be contrary to the interests of the tribe, excites its unvarying disapprobation and is held to be immoral; while every act, on the other hand, which is, as a rule, beneficial to the tribe, is warmly and constantly approved, and is thus considered to be right or moral. From the mental struggle, when an act that would benefit self is injurious to the tribe, there arises conscience; and thus the social instincts are the foundation of the moral sense and of the fundamental principles of morality.[12]

The question of the origin and nature of the moral sense and of conscience is far too vast and complex to be discussed here, and a reference to it has been introduced only to complete the sketch of Mr. Darwin's view of the continuity and gradual development of all human faculties from the lower animals up to savages, and from savage up to civilised man. The point to which I wish specially to call attention is, that to prove continuity and the progressive development of the intellectual and moral faculties from animals to man, is not the same as proving that these faculties have been developed by natural selection; and this last is what Mr. Darwin has hardly attempted, although to support his theory it was absolutely essential to prove it. Because man's physical structure has been developed from an animal form by natural selection, it does not necessarily follow that his mental nature, even though developed *pari passu* with it, has been developed by the same causes only.

To illustrate by a physical analogy. Upheaval and depression of land, combined with sub-aerial denudation by wind and frost, rain and rivers, and marine denudation on coast-lines, were long thought to account for all the modelling of the earth's surface not directly due to volcanic action; and in the early editions of Lyell's *Principles of Geology* these are the sole causes appealed to. But when the action of glaciers was studied and the recent occurrence of a glacial epoch demonstrated as a fact, many phenomena--such as moraines and other gravel deposits, boulder clay, erratic boulders, grooved and rounded rocks, and Alpine lake basins--were seen to be due to this altogether distinct cause. There was no breach of continuity, no sudden catastrophe; the cold period came on and passed away in the most gradual manner, and its effects often passed insensibly into those produced by denudation or upheaval; yet none the less a new agency appeared at a definite time, and new effects were produced which, though continuous with preceding effects, were not due to the same causes. It is not, therefore, to be assumed, without proof or against independent evidence, that the later stages of an apparently continuous development are necessarily due to the same causes only as the earlier stages. Applying this argument to the case of man's intellectual and moral nature, I propose to show that certain definite portions of it could not have been developed by variation and natural selection alone, and that, therefore, some other influence, law, or agency is required to account for them. If this can be clearly shown for any one or more of the special faculties of intellectual man, we shall be justified in assuming that the same unknown cause or power may have had a much wider influence, and may have profoundly influenced the whole course of his development.

The Origin of the Mathematical Faculty.

We have ample evidence that, in all the lower races of man, what may be termed the mathematical faculty is, either absent, or, if present, quite unexercised. The Bushmen and the Brazilian Wood-Indians are said not to count beyond two. Many Australian tribes only have words for one and two, which are combined to make three, four, five, or six, beyond which they do not count. The Damaras of South Africa only count to three; and Mr. Galton gives a curious description of how one of them was hopelessly puzzled when he had sold two sheep for two sticks of tobacco each, and received four sticks in payment. He could only find out that he was correctly paid by taking two sticks and then giving one sheep, then receiving two sticks more and giving the other sheep. Even the comparatively intellectual Zulus can only count up to ten by using the hands and fingers. The Ahts of North-West America count in nearly the same manner, and most of the tribes of South America are no further advanced.[13] The Kaffirs have great herds of cattle, and if one is lost they miss it immediately, but this is not by counting, but by noticing the absence of one they know; just as in a large family or a school a boy is missed without going through the process of counting. Somewhat higher races, as the Esquimaux, can count up to twenty by using the hands and the feet; and other races get even further than this by saying "one man" for twenty, "two men" for forty, and so on, equivalent to our rural mode of reckoning by scores. From the fact that so many of the existing savage races can only count to four or five, Sir John Lubbock thinks it improbable that our earliest ancestors could have counted as high as ten.[14]

When we turn to the more civilised races, we find the use of numbers and the art of counting greatly extended. Even the Tongas of the South Sea islands are said to have been able to count as high as 100,000. But mere counting does not imply either the possession or the use of anything that can be really called the mathematical faculty, the exercise of which in any broad sense has only been possible since the introduction of the decimal notation. The Greeks, the Romans, the Egyptians, the Jews, and the Chinese had all such cumbrous systems, that anything like a science of arithmetic, beyond very simple operations, was impossible; and the Roman system, by which the year 1888 would be written MDCCCLXXXVIII, was that in common use in Europe down to the fourteenth or fifteenth centuries, and even much later in some places. Algebra, which was invented by the Hindoos, from whom also came the decimal notation, was not introduced into Europe till the thirteenth century, although the Greeks had some acquaintance with it; and it reached Western Europe from Italy only in the sixteenth century.[15] It was, no doubt, owing to the absence of a sound system of numeration that the mathematical talent of the Greeks was directed chiefly to geometry, in which science Euclid, Archimedes, and others made such brilliant discoveries. It is, however, during the last three centuries only that the civilised world appears to have become conscious of the possession of a marvellous faculty which, when supplied with the necessary tools in the decimal notation, the elements of algebra and geometry, and the power of rapidly communicating discoveries and ideas by the art of printing, has developed to an extent, the full grandeur of which can be appreciated only by those who have devoted some time (even if unsuccessfully) to the study.

The facts now set forth as to the almost total absence of mathematical faculty in savages and its wonderful development in quite recent times, are exceedingly suggestive, and in regard to them we are limited to two possible theories. Either prehistoric and savage man did not possess this faculty at all (or only in its merest rudiments); or they did possess it, but had neither the means nor the incitements for its exercise. In the former case we have to ask by what means has this faculty been so rapidly developed in all civilised races, many of which a few centuries back were, in this respect, almost savages themselves; while in the latter case the difficulty is still greater, for we have to assume the existence of a faculty which had never been used either by the supposed possessors of it or by their ancestors.

Let us take, then, the least difficult supposition--that savages possessed only the mere rudiments of the faculty, such as their ability to count, sometimes up to ten, but with an utter inability to perform the very simplest processes of arithmetic or of geometry--and inquire how this rudimentary faculty became rapidly developed into that of a Newton, a La Place, a Gauss, or a Cayley. We will admit that there is every possible gradation between these extremes, and that there has been perfect continuity in the development of the faculty; but we ask, What motive power caused its development?

It must be remembered we are here dealing solely with the capability of the Darwinian theory to account for the origin of the *mind*, as well as it accounts for the origin of the *body* of man, and we must, therefore, recall the essential features of that theory. These are, the preservation of useful variations in the struggle for life; that no creature can be improved beyond its necessities for the time being; that the law acts by life and death, and by the survival of the fittest. We have to ask, therefore, what relation the successive stages of improvement of the mathematical faculty had to the life or death of its possessors; to the struggles of tribe with tribe, or nation with nation; or to the ultimate survival of one race and the extinction of another. If it cannot possibly have had any such effects, then it cannot have been produced by natural selection.

It is evident that in the struggles of savage man with the elements and with wild beasts, or of tribe with tribe, this faculty can have had no influence. It had nothing to do with the early migrations of man, or with the conquest and extermination of weaker by more powerful peoples. The Greeks did not successfully resist the Persian invaders by any aid from their few mathematicians, but by military training, patriotism, and self-sacrifice. The barbarous conquerors of the East, Timurlane and Gengkhis Khan, did not owe their success to any superiority of intellect or of mathematical faculty in themselves or their followers. Even if the great conquests of the Romans were, in part, due to their systematic military organisation, and to their skill in making roads and encampments, which may, perhaps, be imputed to some exercise of the mathematical faculty, that did not prevent them from being conquered in turn by barbarians, in whom it was almost entirely absent. And if we take the most civilised peoples of the ancient world--the Hindoos, the Arabs, the Greeks, and the Romans, all of whom had some amount of mathematical talent--we find that it is not these, but the descendants of the barbarians of those days--the Celts, the Teutons, and the Slavs--who have proved themselves the fittest to survive in the great struggle of races, although we cannot trace their steadily growing success during past centuries either to the possession of any exceptional mathematical faculty or to its exercise. They have indeed proved themselves, to-day, to be possessed of a marvellous endowment of the mathematical faculty; but their success at home and abroad, as colonists or as conquerors, as individuals or as nations, can in no way be traced to this faculty, since they were almost the last who devoted themselves to its exercise. We conclude, then, that the present gigantic development of the mathematical faculty is wholly unexplained by the theory of natural selection, and must be due to some altogether distinct cause.

The Origin of the Musical and Artistic Faculties.

These distinctively human faculties follow very closely the lines of the mathematical faculty in their progressive development, and serve to enforce the same argument. Among the lower savages music, as we understand it, hardly exists, though they all delight in rude musical sounds, as of drums, tom-toms, or gongs; and they also sing in monotonous chants. Almost exactly as they advance in general intellect, and in the arts of social life, their appreciation of music appears to rise in proportion; and we find among them rude stringed instruments and whistles, till, in Java, we have regular bands of skilled performers probably the successors of Hindoo musicians of the age before the Mahometan conquest. The Egyptians are believed to have been the earliest musicians, and from them the Jews and the Greeks, no doubt, derived their knowledge of the art; but it seems to be admitted that neither the latter nor the Romans knew anything of harmony or of the essential features of modern music.[16] Till the fifteenth century little progress appears to have been made in the science or the practice of music; but since that era it has advanced with marvellous rapidity, its progress being curiously parallel with that of mathematics, inasmuch as great musical geniuses appeared suddenly among different nations, equal in their possession of this special faculty to any that have since arisen.

As with the mathematical, so with the musical faculty, it is impossible to trace any connection between its possession and survival in the struggle for existence. It seems to have arisen as a *result* of social and intellectual advancement, not as a *cause*; and there is some evidence that it is latent in the lower races, since under European training native military bands have been formed in many parts of the world, which have been able to perform creditably the best modern music.

The artistic faculty has run a somewhat different course, though analogous to that of the faculties already discussed. Most savages exhibit some rudiments of it, either in drawing or carving human or animal figures; but, almost without exception, these figures are rude and such as would be executed by the ordinary inartistic child. In fact, modern savages are, in this respect hardly equal to those prehistoric men who represented the mammoth and the reindeer on pieces of horn or bone. With any advance in the arts of social life, we have a corresponding advance in artistic skill and taste, rising very high in the art of Japan and India, but culminating in the marvellous sculpture of the best period of Grecian history. In the Middle Ages art was chiefly manifested in ecclesiastical architecture and the illumination of manuscripts, but from the thirteenth to the fifteenth centuries pictorial art revived in Italy and attained to a degree of perfection which has never been surpassed. This revival was followed closely by the schools of Germany, the Netherlands, Spain, France, and England, showing that the true artistic faculty belonged to no one nation, but was fairly distributed among the various European races.

These several developments of the artistic faculty, whether manifested in sculpture, painting, or architecture, are evidently outgrowths of the human intellect which have no immediate influence on the survival of individuals or of tribes, or on the success of nations in their struggles for supremacy or for existence. The glorious art of Greece did not prevent the nation from falling under the sway of the less advanced Roman; while we ourselves, among whom art was the latest to arise, have taken the lead in the colonisation of the world, thus proving our mixed race to be the fittest to survive.

Independent Proof that the Mathematical, Musical, and Artistic Faculties have not been Developed under the Law of Natural Selection.

The law of Natural Selection or the survival of the fittest is, as its name implies, a rigid law, which acts by the life or death of the individuals submitted to its action. From its very nature it can act only on useful or hurtful characteristics, eliminating the latter and keeping up the former to a fairly general level of efficiency. Hence it necessarily follows that the characters developed by its means will be present in all the individuals of a species, and, though varying, will not vary very widely from a common standard. The amount of variation we found, in our third chapter, to be about one-fifth or one-sixth of the mean value--that is, if the mean value were taken at 100, the variations would reach from 80 to 120, or somewhat more, if very large numbers were compared. In accordance with this law we find, that all those characters in man which were certainly essential to him during his early stages of development, exist in all savages with some approach to equality. In the speed of running, in bodily strength, in skill with weapons, in acuteness of vision, or in power of following a trail, all are fairly proficient, and the differences of endowment do not probably exceed the limits of variation in animals above referred to. So, in animal instinct or intelligence, we find the same general level of development. Every wren makes a fairly good nest like its fellows; every fox has an average amount of the sagacity of its race; while all the higher birds and mammals have the necessary affections and instincts needful for the protection and bringing-up of their offspring.

But in those specially developed faculties of civilised man which we have been considering, the case is very different. They exist only in a small proportion of individuals, while the difference of capacity between these favoured individuals and the average of mankind is enormous. Taking first the mathematical faculty, probably fewer than one in a hundred really possess it, the great bulk of the population having no natural ability for the study, or feeling the slightest interest

in it.[17] And if we attempt to measure the amount of variation
in the faculty itself between a first-class mathematician and
the ordinary run of people who find any kind of calculation
confusing and altogether devoid of interest, it is probable
that the former could not be estimated at less than a hundred
times the latter, and perhaps a thousand times would more
nearly measure the difference between them.

The artistic faculty appears to agree pretty closely with
the mathematical in its frequency. The boys and girls who,
going beyond the mere conventional designs of children,
draw what they *see*, not what they *know* to be the shape of
things; who naturally sketch in perspective, because it is thus
they see objects; who see, and represent in their sketches, the
light and shade as well as the mere outlines of objects; and
who can draw recognisable sketches of every one they know,
are certainly very few compared with those who are totally
incapable of anything of the kind. From some inquiries
I have made in schools, and from my own observation, I
believe that those who are endowed with this natural artistic
talent do not exceed, even if they come up to, one per cent
of the whole population.

The variations in the amount of artistic faculty are
certainly very great, even if we do not take the extremes. The
gradations of power between the ordinary man or woman
"who does not draw," and whose attempts at representing
any object, animate or inanimate, would be laughable,

and the average good artist who, with a few bold strokes, can produce a recognisable and even effective sketch of a landscape, a street, or an animal, are very numerous; and we can hardly measure the difference between them at less than fifty or a hundred fold.

The musical faculty is undoubtedly, in its lower forms, less uncommon than either of the preceding, but it still differs essentially from the necessary or useful faculties in that it is almost entirely wanting in one-half even of civilised men. For every person who draws, as it were instinctively, there are probably five or ten who sing or play without having been taught and from mere innate love and perception of melody and harmony.[18] On the other band, there are probably about as many who seem absolutely deficient in musical perception, who take little pleasure in it, who cannot perceive discords or remember tunes, and who could not learn to sing or play with any amount of study. The gradations, too, are here quite as great as in mathematics or pictorial art, and the special faculty of the great musical composer must be reckoned many hundreds or perhaps thousands of times greater than that of the ordinary "unmusical" person above referred to.

It appears then, that, both on account of the limited number of persons gifted with the mathematical, the artistic, or the musical faculty, as well as from the enormous variations in its development, these mental powers differ widely from those which are essential to man, and are, for the most part, common to him and the lower animals; and that they could not, therefore, possibly have been developed in him by means of the law of natural selection.

We have thus shown, by two distinct lines of argument, that faculties are developed in civilised man which, both in their mode of origin, their function, and their variations, are altogether distinct from those other characters and faculties which are essential to him, and which have been brought to their actual state of efficiency by the necessities of his existence. And besides the three which have been specially referred to, there are others which evidently belong to the same class. Such is the metaphysical faculty, which enables us to form abstract conceptions of a kind the most remote from all practical applications, to discuss the ultimate causes of things, the nature and qualities of matter, motion, and force, of space and time, of cause and effect, of will and conscience. Speculations on these abstract and difficult questions are impossible to savages, who seem to have no mental faculty enabling them to grasp the essential ideas or conceptions; yet whenever any race attains to civilisation, and comprises a body of people who, whether as priests or philosophers, are relieved from the necessity of labour or of taking an active part in war or government, the metaphysical faculty appears to spring suddenly into existence, although, like the other faculties we have referred to, it is always confined to a very limited proportion of the population.

In the same class we may place the peculiar faculty of wit and humour, an altogether natural gift whose development appears to be parallel with that of the other exceptional faculties. Like them, it is almost unknown among savages, but appears more or less frequently as civilisation advances and the interests of life become more numerous and more complex. Like them, too, it is altogether removed from utility in the struggle for life, and appears sporadically in a very small percentage of the population; the majority being, as is well known, totally unable to say a witty thing or make a pun even to save their lives.[19]

The Interpretation of the Facts.

The facts now set forth prove the existence of a number of mental faculties which either do not exist at all or exist in a very rudimentary condition in savages, but appear almost suddenly and in perfect development in the higher civilised races. These same faculties are further characterised by their sporadic character, being well developed only in a very small proportion of the community; and by the enormous amount of variation in their development, the higher manifestations of them being many times--perhaps a hundred or a thousand times--stronger than the lower. Each of these characteristics is totally inconsistent with any action of the law of natural selection in the production of the faculties referred to; and the facts, taken in their entirety, compel us to recognise some origin for them wholly distinct from that which has served to account for the animal characteristics--whether bodily or mental--of man.

The special faculties we have been discussing clearly point to the existence in man of something which he has not derived from his animal progenitors--something which we may best refer to as being of a spiritual essence or nature, capable of progressive development under favourable conditions. On the hypothesis of this spiritual nature, superadded to the animal nature of man, we are able to understand much that is otherwise mysterious or unintelligible in regard to him, especially the enormous influence of ideas, principles, and beliefs over his whole life and actions. Thus alone we can understand the constancy of the martyr, the unselfishness of the philanthropist, the devotion of the patriot, the enthusiasm of the artist, and the resolute and persevering search of the scientific worker after nature's secrets. Thus we may perceive that the love of truth, the delight in beauty, the passion for justice, and the thrill of exultation with which we hear of any act of courageous self-sacrifice, are the workings within us of a higher nature which has not been developed by means of the struggle for material existence.

It will, no doubt, be urged that the admitted continuity of man's progress from the brute does not admit of the introduction of new causes, and that we have no evidence of the sudden change of nature which such introduction would bring about. The fallacy as to new causes involving any breach of continuity, or any sudden or abrupt change, in the effects, has already been shown; but we will further point out that there are at least three stages in the development of the organic world when some new cause or power must necessarily have come into action.

The first stage is the change from inorganic to organic, when the earliest vegetable cell, or the living protoplasm out of which it arose, first appeared. This is often imputed to a mere increase of complexity of chemical compounds; but increase of complexity, with consequent instability, even if we admit that it may have produced protoplasm as a chemical compound, could certainly not have produced *living* protoplasm--protoplasm which has the power of growth and of reproduction, and of that continuous process of development which has resulted in the marvellous variety and complex organisation of the whole vegetable kingdom. There is in all this something quite beyond and apart from chemical changes, however complex; and it has been well said that the first vegetable cell was a new thing in the world, possessing altogether new powers--that of extracting and fixing carbon from the carbon-dioxide of the atmosphere, that of indefinite reproduction, and, still more marvellous, the power of variation and of reproducing those variations till endless complications of structure and varieties of form have been the result. Here, then, we have indications of a new power at work, which we may term *vitality*, since it gives to certain forms of matter all those characters and properties which constitute Life.

The next stage is still more marvellous, still more completely beyond all possibility of explanation by matter, its laws and forces. It is the introduction of sensation or consciousness, constituting the fundamental distinction between the animal and vegetable kingdoms. Here all idea of mere complication of structure producing the result is out of the question. We feel it to be altogether preposterous to assume that at a certain stage of complexity of atomic constitution, and as a necessary result of that complexity alone, an *ego* should start into existence, a thing that *feels*, that is *conscious* of its own existence. Here we have the certainty that something new has arisen, a being whose nascent consciousness has gone on increasing in power and definiteness till it has culminated in the higher animals. No verbal explanation or attempt at explanation--such as the statement that life is the result of the molecular forces of the protoplasm, or that the whole existing organic universe from the amœba up to man was latent in the fire-mist from which the solar system was developed--can afford any mental satisfaction, or help us in any way to a solution of the mystery.

The third stage is, as we have seen, the existence in man of a number of his most characteristic and noblest faculties, those which raise him furthest above the brutes and open up possibilities of almost indefinite advancement. These faculties could not possibly have been developed by means of the same laws which have determined the progressive development of the organic world in general, and also of man's physical organism.[20]

These three distinct stages of progress from the inorganic world of matter and motion up to man, point clearly to an unseen universe--to a world of spirit, to which the world of matter is altogether subordinate. To this spiritual world we may refer the marvellously complex forces which we know as gravitation, cohesion, chemical force, radiant force, and electricity, without which the material universe could not exist for a moment in its present form, and perhaps not at all, since without these forces, and perhaps others which may be termed atomic, it is doubtful whether matter itself could have any existence. And still more surely can we refer to it those progressive manifestations of Life in the vegetable, the animal, and man--which we may classify as unconscious, conscious, and intellectual life,--and which probably depend upon different degrees of spiritual influx. I have already shown that this involves no necessary infraction of the law of continuity in physical or mental evolution; whence it follows that any difficulty we may find in discriminating the inorganic from the organic, the lower vegetable from the lower animal organisms, or the higher animals from the lowest types of man, has no bearing at all upon the question. This is to be decided by showing that a change in essential nature (due, probably, to causes of a higher order than those of the material universe) took place at the several stages of progress which I have indicated; a change which may be none the less real because absolutely imperceptible at its point of origin, as is the change that takes place in the curve in which a body is moving when the application of some new force causes the curve to be slightly altered.

Concluding Remarks.

Those who admit my interpretation of the evidence now adduced--strictly scientific evidence in its appeal to facts which are clearly what ought *not* to be on the materialistic theory--will be able to accept the spiritual nature of man, as not in any way inconsistent with the theory of evolution, but as dependent on those fundamental laws and causes which furnish the very materials for evolution to work with. They will also be relieved from the crushing mental burthen imposed upon those who--maintaining that we, in common with the rest of nature, are but products of the blind eternal forces of the universe, and believing also that the time must come when the sun will lose his heat and all life on the earth necessarily cease--have to contemplate a not very distant future in which all this glorious earth--which for untold millions of years has been slowly developing forms of life and beauty to culminate at last in man--shall be as if it had never existed; who are compelled to suppose that all the slow growths of our race struggling towards a higher life, all the agony of martyrs, all the groans of victims, all the evil and misery and undeserved suffering of the ages, all the struggles for freedom, all the efforts towards justice, all the aspirations for virtue and the wellbeing of humanity, shall absolutely vanish, and, "like the baseless fabric of a vision, leave not a wrack behind."

As contrasted with this hopeless and soul-deadening belief, we, who accept the existence of a spiritual world, can look upon the universe as a grand consistent whole adapted in all its parts to the development of spiritual beings capable of indefinite life and perfectibility. To us, the whole purpose, the only *raison d'être* of the world--with all its complexities of physical structure, with its grand geological progress, the slow evolution of the vegetable and animal kingdoms, and the ultimate appearance of man--was the development of the human spirit in association with the human body. From the fact that the spirit of man--the man himself--*is* so developed, we may well believe that this is the only, or at least the best, way for its development; and we may even see in what is usually termed "evil" on the earth, one of the most efficient means of its growth. For we know that the noblest faculties of man are strengthened and perfected by struggle and effort; it is by unceasing warfare against physical evils and in the midst of difficulty and danger that energy, courage, self-reliance, and industry have become the common qualities of the northern races; it is by the battle with moral evil in all its hydra-headed forms, that the still nobler qualities of justice and mercy and humanity and self-sacrifice have been steadily increasing in the world. Beings thus trained and strengthened by their surroundings, and possessing latent faculties capable of such noble development, are surely destined for a higher and more permanent existence; and we may confidently believe with our greatest living poet--

That life is not as idle ore,
But iron dug from central gloom,
And heated hot with burning fears,
And dipt in baths of hissing tears,
And batter'd with the shocks of doom
To shape and use.

We thus find that the Darwinian theory, even when carried out to its extreme logical conclusion, not only does not oppose, but lends a decided support to, a belief in the spiritual nature of man. It shows us how man's body may have been developed from that of a lower animal form under the law of natural selection; but it also teaches us that we possess intellectual and moral faculties which could not have been so developed, but must have had another origin; and for this origin we can only find an adequate cause in the unseen universe of Spirit.

Notes Appearing in the Original Work

1. *Descent of Man*, pp. 41-43; also pp. 13-15.

2. *Man's Place in Nature*, p. 64.

3. *Man's Place in Nature*, p. 67. See Figs. of Embryos of Man and Dog in Darwin's *Descent of Man*, p. 10.

4. *The Descent of Man*, pp. 7, 8.

5. *Man and Apes*. By St. George Mivart, F.R.S., 1873. It is an interesting fact (for which I am indebted to Mr. E. B. Poulton) that the human embryo possesses the extra rib and wrist-bone referred to above in (2) and (4) as occurring in some of the apes.

6. *Man and Apes*, pp. 138, 144.

7. For a sketch of the evidence of Man's Antiquity in America, see *The Nineteenth Century* for November 1887.

8. This subject was first discussed in an article in the *Anthropological Review*, May 1864, and republished in my *Contributions to Natural Selection*, chap. ix, in 1870.

9. *Man's Place in Nature*, p. 102.

10. For a full discussion of this question, see the author's *Geographical Distribution of Animals*, vol. I. p. 285.

11. For a full discussion of all these points, see *Descent*

of Man, chap. iii.

12. *Descent of Man*, chap. iv.

13. Lubbock's *Origin of Civilisation*, fourth edition, pp. 434-440; Tylor's *Primitive Culture*, chap. vii.

14. It has been recently stated that some of these facts are erroneous, and that some Australians can keep accurate reckoning up to 100, or more, when required. But this does not alter the general fact that many low races, including the Australians, have no words for high numbers and never require to use them. If they are now, with a little practice, able to count much higher, this indicates the possession of a faculty which could not have been developed under the law of utility only, since the absence of words for such high numbers shows that they were neither used nor required.

15. Article Arithmetic in *Eng. Cyc. of Arts and Sciences*.

16. See "History of Music," in *Eng. Cyc.*, Science and Arts Division.

17. This is the estimate furnished me by two mathematical masters in one of our great public schools of the proportion of boys who have any special taste or capacity for mathematical studies. Many more, of course, can be drilled into a fair knowledge of elementary mathematics, but only this small proportion possess the natural faculty which renders it possible for them ever to rank high as mathematicians, to take any pleasure in it, or to do any original mathematical work.

18. I am informed, however, by a music master in a large school that only about one per cent have real or decided musical talent, corresponding curiously with the estimate of the mathematicians.

19. In the latter part of his essay on Heredity (pp. 91-93 of the volume of *Essays*), Dr. Weismann refers to this question of the origin of "talents" in man, and, like myself, comes to the conclusion that they could not be developed under the law of natural selection. He says: "It may be objected that, in man, in addition to the instincts inherent in every individual, special individual predispositions are also found, of such a nature that it is impossible they can have arisen by individual variations of the germ-plasm. On the other hand, these predispositions--which we call talents--cannot have arisen through natural selection, because life is in no way dependent on their presence, and there seems to be no way of explaining their origin except by an assumption of the summation of the skill attained by exercise in the course of each single life. In this case, therefore, we seem at first sight to be compelled to accept the transmission of acquired characters." Weismann then goes on to show that the facts do not support this view; that the mathematical, musical, or artistic faculties often appear suddenly in a family whose other members and ancestors were in no way distinguished; and that even when hereditary in families, the talent often appears at its maximum at the commencement or in the

middle of the series, not increasing to the end, as it should do if it depended in any way on the transmission of acquired skill. Gauss was not the son of a mathematician, nor Handel of a musician, nor Titian of a painter, and there is no proof of any special talent in the ancestors of these men of genius, who at once developed the most marvellous pre-eminence in their respective talents. And after showing that such great men only appear at certain stages of human development, and that two or more of the special talents are not unfrequently combined in one individual, he concludes thus--"Upon this subject I only wish to add that, in my opinion, talents do not appear to depend upon the improvement of any special mental quality by continued practice, but they are the expression, and to a certain extent the bye-product, of the human mind, which is so highly developed in all directions." It will, I think, be admitted that this view hardly accounts for the existence of the highly peculiar human faculties in question.

20. For an earlier discussion of this subject, with some wider applications, see the author's *Contributions to the Theory of Natural Selection*, chap. x.

www.ingramcontent.com/pod-product-compliance
Lightning Source LLC
Chambersburg PA
CBHW021639270326
41931CB00008B/1077